Islands

Catherine Chambers

Heinemann Library

Designed by David Oakley
Illustrations by Tokay Interactive
Originated by Dot Gradations
Printed in Hong Kong/China

06 05 04 03 02
10 9 8 7 6 5 4 3 2 1

Library of Congress Cataloging-in-Publication Data
Chambers, Catherine, 1954-
   Islands / Catherine Chambers.
      p. cm. – (Mapping earthforms)
   Includes bibliographical references (p.) and index.
   Summary: Explores the world's islands, discussing how they were formed, what organisms live there, and how they are used by humans.
   ISBN 1-57572-523-1 (lib. bdg.)   ISBN 1-4034-0033-4 (pbk. bdg.)
   1. Islands—Juvenile literature. [1. Islands.] I. Title.
GB471.C48 2000
508.14'2 21—dc21
                                               99-043373

Acknowledgments
The Publishers would like to thank the following for permission to reproduce photographs: Science Photo Library/U.S. Geological Survey, p. 4; Harding Picture Library, p. 5; Still Pictures/E. Hussenet, p. 6; Still Pictures/ G. and M. Moss, p. 8; Robert Harding Picture Library/N. Wheeler, p. 10; Robert Harding Picture Library/ R. Richardson, p. 11; Robert Harding Picture Library/G. Williams, p. 13; Empics/T. Marshall, p. 14; Still Pictures/H. Klein, p. 16; Still Pictures/B. Brecelj, p. 17; Bruce Coleman Limited/K. Wothe, p. 18; Bruce Coleman Limited/J. Cancalosi, p. 19; Robert Robert Harding Picture Library/F. Hall, p. 20; Corbis, p. 21; Robert Harding Picture Library/P. van Riel, p. 23, Still Pictures/M. Edwards, p. 24; Oxford Scientific Films/G. Soury, p. 26; Robert Harding Picture Library/L. Wilson, p. 27; Oxford Scientific Films/H. Bardarson, p. 29.

Cover photograph reproduced with permission of Robert Harding Picture Library.

Every effort has been made to contact copyright holders of any material reproduced in this book. Any omissions will be rectified in subsequent printings if notice is given to the publisher.

Some words are shown in bold, **like this.** You can find out what they mean by looking in the glossary.

Contents

What Is an Island?

An island is a land mass that is completely surrounded by water. Some islands are so tiny that you could easily walk all the way around them. Others are hundreds of miles long. Most islands are in the ocean near the coasts of the continents. Some islands rise in the middle of the ocean. Islands are also in rivers, lakes, and **deltas** where rivers meet the ocean. We will look at different types and sizes of islands all around the world.

How are islands formed?

Some islands are the peaks of volcanoes that rise from the ocean floor. Other islands were formed when the sea level rose over the bottoms of mountains, leaving high peaks sticking out of the water. A river can carve a course around an island of hard rock.

A group of islands is called an **archipelago**. The islands in the archipelago are usually formed in the same way, but the plants and flowers on each island can be quite different. This archipelago is the state of Hawaii. It is made up of eight main islands and over one hundred smaller ones.

Hard rock can also form an island in a lake. A slow-moving river might split near the ocean and wind around to make islands of fine soil.

Many islands are warm and tropical like this one, but many other islands are covered in a huge cap or sheet of ice.

What do islands look like?

Small island landscapes range from hot **coral** sands to cold, windswept rocks. Large islands can have a variety of landscapes, from mountains and **plateaus** to flat river plains. We will discover how size and **climate** affect an island landscape.

Life on an island

Islands can be **isolated**. Sometimes special **species** of plants and animals have **evolved** on them. But what about people? Today, most island communities have contact with the people on large continents. They are able to buy many goods that they cannot produce on their own. Some island people use the natural materials around them to build their homes and make a living. What does the future hold for life on the islands of the world?

Islands of the World

Where are the islands?

There are thousands of islands all over the world. Most of them lie near the continents, but some are farther out in the oceans and seas. Other islands form in the middles of lakes, rivers, and **deltas**. You cannot see these islands on the map to the right, because they are too small.

It is sometimes difficult to decide what is an island and what is a continent. Greenland is called an island. It is the largest island in the world. It is just about 840,000 square miles (over 2 million square kilometers). Australia is called a continent, even though it is surrounded by water. It is more than three times as big as Greenland—too big to be an island. Australia is so large that it has several different **climates** and types of vegetation. Most big islands have a range of temperatures and rainfall, but the climate is not as varied. Smaller islands do not have as much variety.

Canada has more islands than any other country. It also has many of the biggest, including Baffin Island and Ellesmere Island. The Arctic **archipelago** in the north includes 31 islands with an area of more than 500 sq. mi. (1,300 sq km). Most of Canada's islands are along its long coastline. These islands are home to many different plants and animals, such as polar bears. These inhabitants have had to adjust to the extreme temperatures and conditions of the Arctic.

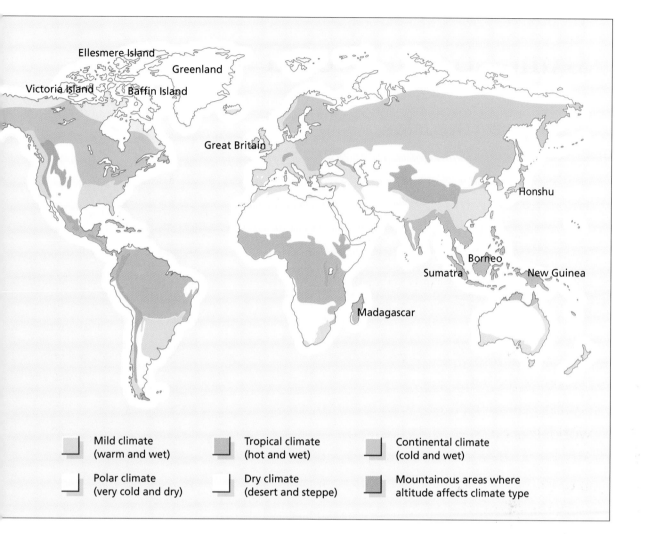

Mild climate
(warm and wet)

Tropical climate
(hot and wet)

Continental climate
(cold and wet)

Polar climate
(very cold and dry)

Dry climate
(desert and steppe)

Mountainous areas where
altitude affects climate type

Islands in different climates

There are many islands in the cold northern part of the world from Canada to Russia. Most of these islands were formed millions of years ago.

Small **coral** creatures live in great numbers in the warmest seas. Here are the world's coral islands, which are made up of dead and living coral. Volcanic islands are not only found where the climate above ground is hot. They are also formed where the crust of the earth under the ocean is weak and very hot.

Islands are scattered all over the world. There are many in the middle of the Pacific Ocean. Many of these islands were formed by volcanoes erupting from the ocean floor. Some are too small to appear on this map. There are fewer islands in the middle of the Atlantic Ocean.

How Are Islands Formed?

Islands in the ocean

Some islands were once mountains on the edges of continents. During the last **ice age** they were covered in snow and ice. When the **climate** began to warm up, about 11,000 years ago, the ice and snow melted and the sea rose. The waters covered the mountains on the coasts, leaving just the peaks sticking out. Norway has 3,000 islands like this around its shores.

Other islands were formed when the sea filled a low spot between a small piece of land and a continent. The low spot was caused by a strip of land that slipped under huge cracks in the earth's crust. The British Isles were separated from Europe in this way, and Sicily was **isolated** from Italy.

This ring-shaped island has a lake called a lagoon in the middle. The island was formed by a volcano. In warm waters, **coral** creatures cling to the volcanic rocks. When they die, they leave hard, crusty layers. The original mountain sinks, leaving behind a coral island ring called an **atoll**.

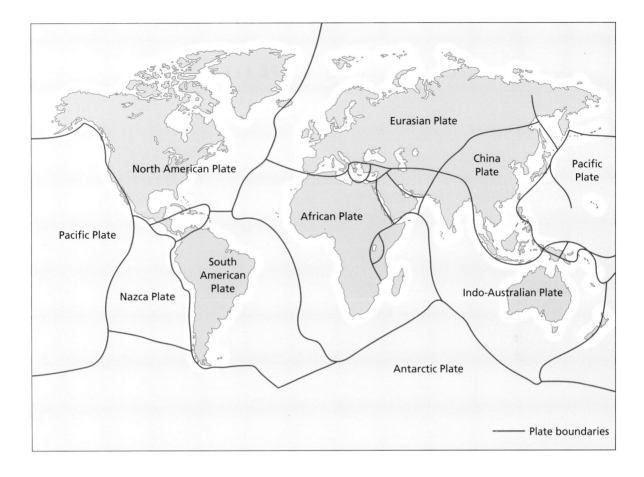

Plate boundaries

Some islands are huge bars of sand pushed up by ocean tides. These islands are called barrier islands. They help protect the shoreline from heavy waves. Fraser Island, off the east coast of Australia, is a barrier island.

Lake and river islands

Some islands rise in the middle of wide rivers that wind their way along flat plains. The river **scours** out its course through soft rock and curves around either side of an island of harder rock. This can happen in the middle of lakes, too. Islands can also form in **deltas**, where the river splits into streams that make their way around islands of thick, muddy **silt**.

Hundreds of islands lie along the bold lines on the map. They mark the edges of **tectonic plates**. The plates sit on top of a layer of hot, sticky rock called the **mantle.** Islands form where the edges of the tectonic plates push together or pull apart, or the islands erupt suddenly through a weak point in the earth's crust. The Azores in the Atlantic Ocean and Hawaii in the Pacific Ocean are volcanic islands.

Island Landscapes

The landscape of an island depends on its origin, its size, and its location. Some big islands have a range of landscapes, while others look nearly the same all over. Some small islands rise as mountains with streams and waterfalls or with hot, volcanic springs. Other islands might be bare rocks with no fresh water.

Large island landscapes

Greenland is the world's largest island. In many ways it is like other islands in the Arctic region. The main island is mostly hard, granite rocks that rise to a high, flat **plateau**. Eighty percent of Greenland is covered in a permanent **ice cap** with just a few peaks sticking out. Tightly curved inlets, called **fjords**, make a fringe all around Greenland's coast. The fjords were formed by **glaciers** carving their way from the upland down to the sea. Icebergs drift around Greenland's shores. Many smaller islands surround the coast.

Guam is a small island that lies in the Pacific Ocean. It has a **coral** island plateau in the north and volcanic hills and valleys to the south. Guam lies at the southern end of a submerged mountain range that is 1,500 miles (2,500 kilometers) long.

The island of Great Britain lies south of Greenland, where it is warmer. Great Britain was separated from the continent of Europe when a strip of land slipped lower, allowing water to flow over it. Great Britain has green hills, flat plains, and just a few mountains. There are many rivers and lakes. Many of the hills and lakes were formed by glaciers during the last **ice age**. Glaciers also left behind mounds and plains of **eroded** rocks, sands, and soils. Great Britain's coastline, though, has mostly been formed by waves battering the rocks.

This beach of black, volcanic sand is on Tenerife in the Canary Islands. Volcanic islands often have rich volcanic soils in which grapes and other vegetables are grown.

Small island landscapes

Some small islands are volcanoes sticking out of the ocean, such as the Canary Islands off the northwest coast of Africa. These volcanic islands have rich, fine volcanic soil on dry, mountainous slopes. Black volcanic sand surrounds the coastline. Some small islands in the Mediterranean Sea are not volcanic, but they are stony, mountainous, and very dry. In wetter parts of the world, streams and rivers help to shape the island landscapes.

Islands and the Weather

The oceans and seas are where strong winds blow and where rain is made. Small, unprotected islands are easily affected. Larger islands often have different weather systems blowing on them at the same time. It can be warm and sunny in one part of the island and cold and wet in another.

Wind and rain over the oceans

As wind blows over the oceans, the air picks up moisture. This moisture is held in the air as drops of **water vapor**, which rise and form clouds. The clouds shed rain when they are blown over an island, especially if

This map shows the main **trade winds** that blow between April and October. Sailors relied on the trade winds to blow their ships to trading ports. The trade winds often meet at islands in the **Tropics**, bringing heavy rain. **Monsoon winds** change direction for half of the year. In September and October, they bring heavy rain to the islands in the western Pacific Ocean.

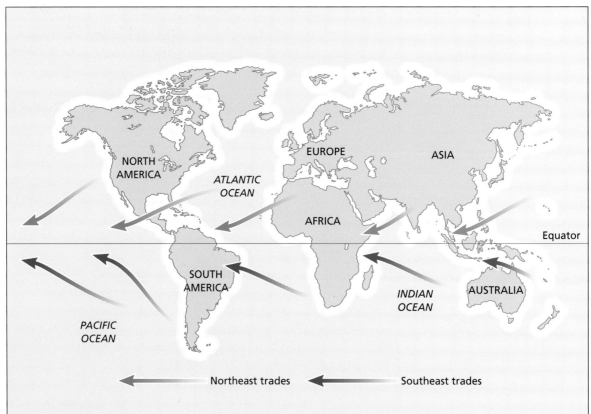

NORTH AMERICA

ATLANTIC OCEAN

EUROPE

ASIA

AFRICA

Equator

SOUTH AMERICA

INDIAN OCEAN

AUSTRALIA

PACIFIC OCEAN

Northeast trades

Southeast trades

the island is mountainous. This is because the cloud rises up the mountainside and the water vapor in the cloud cools and becomes water droplets. The clouds often shed all their rain on one side of the mountain, making the other side quite dry.

There are parts of the world where the winds blow over a huge stretch of ocean and bring a lot of rain to islands in their path. But the Greek Islands in the Mediterranean Sea are very dry, because the winds do not have time to pick up a lot of moisture before they reach the islands.

Winds can also be affected by warm ocean **currents.** The Hebrides Islands, off the northwest coast of Great Britain, are warmed by a current called the North Atlantic Drift. The Hebrides are as far north as very cold Canadian islands, but they are warmer because of the current.

The islands of the Caribbean and the islands off the south coast of the United States are often badly hit by **hurricanes**. They occur when there are especially large amounts of very warm, moist air rising over the ocean. Hurricane winds can sometimes reach 155 miles (250 kilometers) per hour.

Island Life—Honshu

Honshu is Japan's largest island and the seventh largest island in the world. To the west, it is separated from China by the Sea of Japan. The Pacific Ocean stretches away from Honshu's eastern shores. Honshu was once a mountainous coastal strip on the edge of the Asian continent. But the land between the mountains and the continent slipped down and filled with sea water. Over 200 small islands lie around Honshu.

There are many volcanic mountains on Honshu. Some are active. The highest is Mount Fuji, which rises 12,388 feet (3,776 meters) above sea level. Mount Fuji has very beautiful scenery and is a sacred peak for Japanese people. Many rivers begin in the mountains, which also contain lakes. Honshu has three of Japan's largest rivers— the Tone, Shinano, and Kino.

◆ Island life on Honshu varies with the landscape and **climate.** The mountainous northern region is a center for winter sports, especially skiing. The 1998 Winter Olympics were held here, in Nagano. The low-lying coastal regions have a milder climate. Many of the larger cities and factories are here.

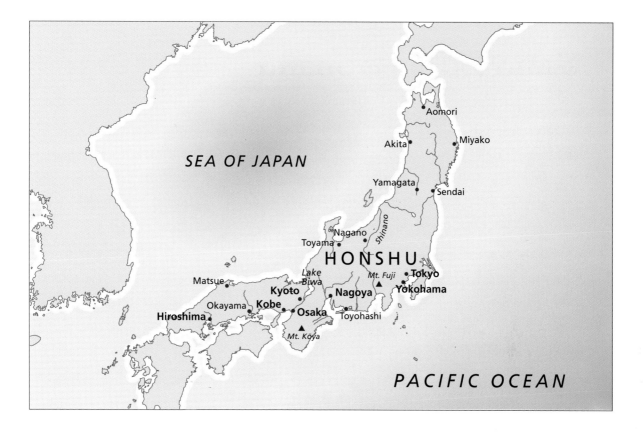

There are many rivers and lakes on Honshu because a lot of rain reaches the island, especially between September and October. This is when the **monsoon winds** from the southwest bring storms and flooding to the island. The winds bring **hurricanes,** too. Hurricanes are called typhoons in the western Pacific.

Summers on Honshu are hot and humid. The summer temperature can reach 95°F (35°C). Winter brings different types of weather to different parts of the island. The west coast is cold, and the mountains are freezing. But the warm Kuro Siwo **current** brings milder temperatures to the east side of Honshu. Most people live on the eastern lowlands. The different temperatures create a variety of **habitats** for many types of plants and animals.

Although Honshu was not formed volcanically, it lies in the Ring of Fire, where there is a lot of volcanic activity. Honshu is home to many hot water springs, as well as active volcanoes. There are also frequent earthquakes.

Island Plants

Islands are located in many different **climates.** They are different sizes and have a variety of landscapes. Some islands are very dry and rocky, while others are hit by fierce rainstorms. Because of this, there are many types of island plants. Some of them have had to **adapt** to extreme conditions, such as strong winds.

Isolated islands

Many islands have plants that can be found in many other areas of the world. These plants came to the islands in a variety of ways. Seeds can be lifted by the wind and blown on to islands. Seeds are also dispersed in bird droppings. Large seed capsules, such as coconuts, can float over the oceans. People can also bring plant **species** to an island.

Arctic islands are frozen, dry, dark, and very windy during the long winters. The plants that grow there are especially adapted for such a harsh climate. Tiny mosses and **lichens** cling to the small patches of bare rock. When the short summer comes, the top layer of soil thaws, and tough grasses and stubby, creeping shrubs grow. **Saxifrages** have tightly packed flowers to protect them from the wind. Low-growing berry bushes hug the ground. This region is called tundra, which means "treeless plain."

Islands are special because they are **isolated**. Islands formed by separation from a continent often have unique **species** of larger plants. Many of these plants originally came from the nearby continents. But over time, they have changed and adapted to conditions on the island. They have developed into subspecies with features of their own.

Volcanic islands, on the other hand, are totally bare when they are first formed. But most have been colonized, or taken over, by plants from other lands. The Hawaiian Islands are volcanic and have very lush vegetation. Many of the Hawaiian plants developed from seeds carried by ocean **currents** and the wind. Fruit-eating **migratory** birds also left seeds behind before flying away.

The island of Jamaica lies in the Caribbean Sea, near Central America. Its landscape is varied, with low-lying plains on the coast and high mountains in the east. As a result, Jamaica has a great variety of plant species, including over 2,000 species of flowering plants. Many flowers are very big and bright to attract the insects and birds that **pollinate** them. Jamaica has many varieties of palms, and it has forests of hardwood trees, such as mahogany and rosewood.

17

Island Creatures

Many islands with lots of vegetation and mild or hot **climates** have a wide variety of animal life. Islands in the **polar regions** are so cold that they have less vegetation and fewer varieties of animals. Most **species** in the polar regions live on the shores of the islands, which are thriving **habitats** for seals and seabirds, such as penguins.

The special thing about island species is that they are often unique. This means that they have **evolved** differently from their relatives on nearby continents, or even on islands very close to them. Animals often have to compete for food, and competition on small islands is very fierce. So each species has evolved to eat a particular type of island vegetation. This way none of them will starve. It also makes sure that one type of food does not get eaten so much that it dies out.

Madagascar is a large island that lies in the Indian Ocean near the east coast of Africa. Like many islands, it does not have any really large mammals. But it does have several unique species of **primates**, called lemurs.

The Galapagos Islands are a group of volcanic islands in the Pacific Ocean. They are about 620 miles (1,000 kilometers) west of Ecuador. Creatures on each of the islands are surprisingly different from one another. Giant tortoises have **adapted** to the contrasting island environments and even look different from each other. The tortoises from the dry islands have developed long necks and high curves in the shell where the neck stretches. This is so that the tortoise can reach up to high cactus branches and spiny cactus leaves. On wetter islands the tortoises have adapted to eating grasses and low-growing plants. These tortoises have shorter necks and a scooped bottom shell so that they can bend their heads down easily.

Tasmania is an island not far from the southern tip of Australia. It is home to the Tasmanian devil, a ferocious, meat-eating **marsupial.** Tasmanian devils are not found anywhere else.

Island People

Since ancient times, communities have sailed from continents to islands in search of a better place to grow crops and raise animals. The long island coasts have provided fish and harbors for boats. Some islands, especially those of volcanic origin, have deposits of gemstones and **minerals**, such as gold. Today, small islands attract people from the mainland who want a simple, peaceful way of life. But young islanders often leave their homes to find work in bigger, busier areas.

The island of Hong Kong lies near the coast of mainland China. The island has very few natural resources, but it is one of the busiest trading centers in the world. This is because it lies in the path of many trade routes and has deep, natural harbors around its coast.

Making a living

Frozen islands, whether big or small, do not have large populations. There is not enough unfrozen land to grow crops or raise animals. Minerals are very difficult to extract from the ground. But some Arctic islands do support communities of Inuit, who use their environment to survive. They fish and hunt caribou, reindeer, seals, whales, and walruses.

Some islands are hot, rocky, mountainous, and poor for farming, except around the coastal lowlands. The Greek Islands are like this, but people have lived there for thousands of years. They have made their living from fishing, shipping, and trading. Now these islands are popular with tourists, as are many of the warm islands around the world.

Volcanic islands have good soils that attract farmers. Some volcanic islands also have hot springs that can provide energy. In Reykjavik, the capital of the island of Iceland, most buildings are heated by hot water piped from **geysers** and hot springs. However, volcanic islands can be dangerous places to live. The island of Montserrat in the Caribbean Sea erupted in 1996, and most of the inhabitants had to leave.

The **isolation** of islands can be useful. For many years, the U.S. controlled **immigration** by making everyone stop at Ellis Island in New York harbor to check in. Islands have also been used as prisons. Alcatraz, a small island in San Francisco Bay, was an "escape-proof" U.S. prison until it closed in 1963.

A Way of Life—The Polynesians

Polynesia is a group of widely scattered islands in the central Pacific Ocean from Hawaii in the north to New Zealand in the south. The ancestors of today's Polynesians came to the islands over 2,000 years ago, probably from the Malay **Peninsula.** They traveled east on large wooden rafts to find new settlements. The people of Polynesia use their natural environment to make a living and build homes.

The islands of Polynesia lie in a triangular area with points at New Zealand, the Hawaiian Islands, and Easter Island. The islands are far from continents and are open to natural disasters, such as **hurricanes** and **tsunamis**.

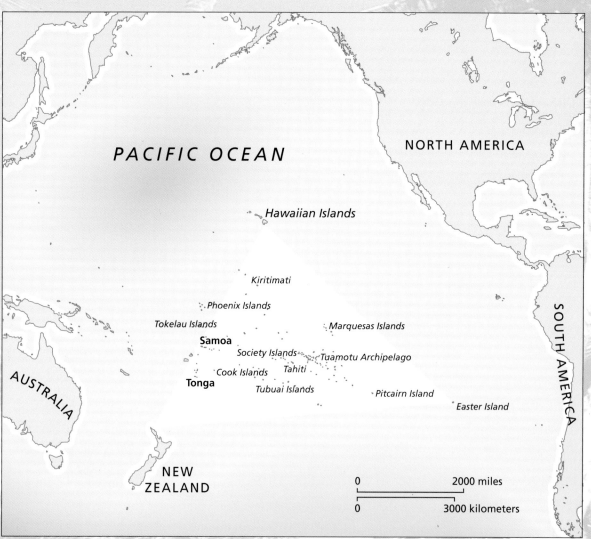

PACIFIC OCEAN

NORTH AMERICA

Hawaiian Islands

Kiritimati

Phoenix Islands

Tokelau Islands

Marquesas Islands

Samoa

Society Islands

Tuamotu Archipelago

Cook Islands

Tahiti

Tonga

Tubuai Islands

Pitcairn Island

Easter Island

AUSTRALIA

SOUTH AMERICA

NEW ZEALAND

0 ———— 2000 miles

0 ———— 3000 kilometers

Building a home

Some Polynesian islands are volcanic, while others are made of limestone or **coral**. There is often lush vegetation with many trees, including palms and precious hardwoods. Traditional houses come in many styles, but most are made from a hardwood post frame with walls of bamboo and palm leaves. The roof is thatched with reeds. Today many people live in modern houses that are built to withstand hurricanes.

The soil on volcanic islands is especially **fertile**. One of the main foods in Polynesia is taro. This is a root crop that is baked, pounded, and made into soft dough. Yams, vegetables, and tropical fruits are also grown.

Coconut palms provide food and more. The fibers on the outer husk are called coir. Coir is made into rope and used to stuff furniture. Coir and leaves are woven into baskets and matting. Coconut oil is used in cooking and to make skin creams and soap.

Fishing is one of the main industries in the Polynesian Islands. Fishing canoes are carved from solid pieces of wood from the hardwood forests. The canoes have beautiful carvings in front. Modern industries include processing and packing the natural plants and fruits of the islands, and mining the gold and other **minerals** found on some of the volcanic islands.

The islands of Polynesia are well-known for their highly decorated carvings and cloth designs. Each island has its own style. Many tourists buy carvings and decorated cloth made from tree bark and other natural fibers.

Changing Islands

Nature causes change

Islands change all the time. Some rise from the ocean floor or sink down under the waves. Others grow as **coral** deposits that build up layers of land. The sea constantly batters the coastline, wearing it away and changing its shape. Inland, strong winds and heavy rain **erode** mountain landscapes and widen river valleys.

Most changes take place over thousands of years, but some take much less time. **Hurricanes** can cause change to happen very quickly. In some parts of the world, coastlines are eroding and cliff-top villages are in danger of falling into the sea. Some islands are sinking. Erosion and sinking are occurring partly because there is more water in the oceans. The **ice caps, glaciers**, and icebergs of the Arctic and Antarctic are melting. This causes sea levels to rise.

Haiti shares the Caribbean island of Hispaniola with another country, the Dominican Republic. Many of Haiti's lush forests have been cut down, leaving the land bare. People, not nature, have caused this change.

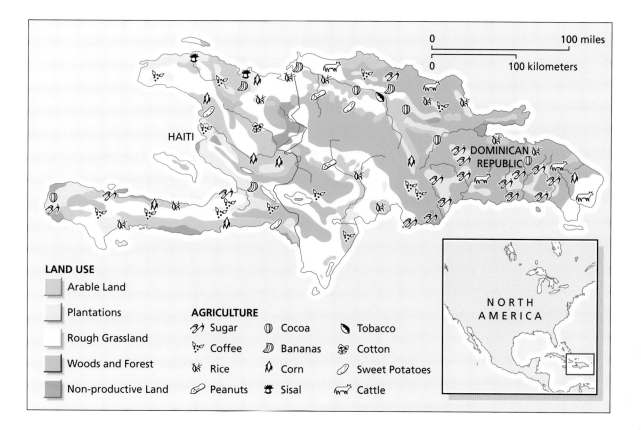

LAND USE
- Arable Land
- Plantations
- Rough Grassland
- Woods and Forest
- Non-productive Land

AGRICULTURE

Sugar	Cocoa	Tobacco
Coffee	Bananas	Cotton
Rice	Corn	Sweet Potatoes
Peanuts	Sisal	Cattle

People cause change

People on the islands also change the life around them. Sometimes they clear away the natural plants to grow crops. Cutting down too many trees has led to bare islands. This creates a poor **habitat** for wildlife and difficult farming conditions.

People can change islands in other ways. Bringing a new **species** of plant or animal to an island can harm native species and upset the delicate balance of life. Some nations have also used islands for testing nuclear weapons. Changes like these can affect an island forever.

Hispaniola is often hit by hurricanes. These can cause landslides on the hillsides, which destroy crops in the fields below. Soil erosion is also a big problem.

25

Looking to the Future

The swelling seas

The future of islands is uncertain. The earth is becoming warmer, which causes sea levels to rise. Some scientists think that temperatures are changing because of the natural cycle of the earth's **climate**. They believe that it is natural for the earth's temperature to rise and fall over time.

Others believe that the earth is getting warmer because of the greenhouse effect. This is caused by carbon dioxide and other gases in the **atmosphere.** The gases trap the sun's rays, and the temperature rises. Greenhouse gases come from sources like

Because island resources are limited, islanders must be careful to use resources that are renewable. The island of Manihi in Polynesia has developed black pearl farms. This way, pearls are made to replace the ones that are sold. Manihi's economy is also helped by tourism. Tourists come because of the climate and the famous diving center.

car exhaust, manufacturing, and power plants. Trees can absorb carbon dioxide, but the number of trees on earth is decreasing. Replanting forests, recycling, and finding ways of using cleaner sources of energy are possible answers.

Support for the islands

Many islands are home to people with unique skills, traditions, and customs. Island **habitats** have special **species** of plants and creatures, too. How can they be protected?

We need to make sure that tourism does not destroy life on the islands. In the Galapagos Islands, only a small number of tourists can visit at any one time. This has protected both islanders and natural island habitats.

The economies of many small islands depend on tourism and the sale of one or two kinds of crop to other parts of the world. But **hurricanes** can destroy hotels, and a disease can wipe out an entire crop. Many islands are trying to develop other industries, so that if one fails, their economy can still survive.

The Hebrides Islands lie off the northwest coast of Great Britain. Many people make their living by raising sheep and making woolen cloth and clothes. But raising sheep takes a lot of land, and the islands are small. So some people set up small computer software or communications businesses away from cities and towns. Others have left their islands to find work in other countries, as they have done for hundreds of years.

27

Island Facts

The top ten islands

These are the ten largest islands in the world. Half of them are in the far northern parts of the world.

	Location	Area (square miles)	(square kilometers)
Greenland	North Atlantic Ocean	840,000	2,175,600
New Guinea	western Pacific Ocean	306,000	821,030
Borneo	western Pacific Ocean	280,100	744,360
Madagascar	western Indian Ocean	226,660	587,040
Baffin Island	Arctic Ocean	196,100	508,000
Sumatra	eastern Indian Ocean	182,860	473,600
Honshu	western Pacific Ocean	87,805	230,500
Great Britain	North Atlantic Ocean	84,200	229,880
Victoria	Arctic Ocean	83,900	212,200
Ellesmere	Arctic Ocean	75,767	212,000

Islands and evolution

Over 160 years ago, the naturalist Charles Darwin studied plants and animals on the Galapagos Islands. He used the information he gathered there to develop his theories about the ways that animals and plants **evolve** separately over time and in different **habitats.**

The coast of Norway is protected by a line of thousands of islands called the Skerry Guard. The islands protect the shores from strong waves, wind, and **erosion**.

In 1963 a new volcanic island erupted off the southwest coast of Iceland. After three years the volcano started to calm down. It is now nearly 660 feet (200 meters) high, and many plants and animals already thrive there. The island was named Surtsey, for the ancient Icelandic god of fire.

Glossary

adapt to change and make suitable for a new use

archipelago group of islands in which the islands have usually all been formed in the same way

atmosphere layer of gases that surrounds the earth

atoll ring-shaped island made of coral

climate rainfall, temperature, and winds that normally affect a large area

coral hard rock made of the shells of tiny dead sea creatures cemented together with limestone made by the creatures themselves

current strong surge of water that flows constantly in one direction

delta where a river meets the sea, and the water spreads out into many smaller streams

erosion wearing away of rocks and soil by wind, water, ice, or acid

evolve when, over a very long time, creatures and plants develop features and habits that help them to survive well in their environment

fertile describes rich soil in which crops can grow easily

fjord deep, narrow inlet of the sea with very steep sides

geyser spring heated by underground volcanic activity that shoots hot water and steam into the air

glacier thick mass of ice formed from compressed snow

habitat place where a plant or animal grows or lives in nature

hurricane storm with winds that blow faster than 75 miles (120 kilometers) per hour

ice age time when snow and ice covered much of the earth. There have been many ice ages in Earth's history, separated by periods of warmer weather.

ice cap large area of ice that mostly covers the land all year

immigration moving permanently from one country to another

isolated alone, cut off from the rest of the world

lichen mixture of a fungus and algae that is not a true plant

mantle layer of hot, molten rock on which the earth's crust sits

marsupial type of mammal that carries its young in a pouch on the mother's belly

migratory moving from one region to another, depending on the season

mineral substance that is formed naturally in rocks and earth, such as coal, tin, or salt

monsoon winds warm, wet summer winds that blow around India and the western Pacific Ocean

peninsula narrow tongue of land jutting out into the sea

plateau area of high, flat ground, often lying between mountains

polar region area around the North and South Poles

pollinate to transfer pollen from one plant to another in order to create fruit and new plants

primate group of creatures including monkeys, lemurs, apes, and humans

saxifrage type of flowering plant that grows in the arctic regions

scour to rub hard against something, wearing it away

silt fine particles of eroded rock and soil that can settle in lakes and rivers, sometimes blocking the movement of water

species one of the groups used for classifying animals. The members of each species are very similar.

tectonic plate area of the earth's crust separated from other plates by deep cracks. Earthquakes, volcanic activity, and the forming of mountains take place where these plates meet

trade wind wind that blows steadily from the east toward the Equator

Tropics the region between the Tropic of Cancer and the Tropic of Capricorn, two imaginary lines drawn around the earth, above and below the Equator.

tsunami huge, destructive wave caused by an undersea earthquake

water vapor water that has been heated so much that it forms a gas that is held in the air. Drops of water form again when the vapor is cooled.

More Books to Read

Hooper, Rosanne. *Life in the Islands.* New York: Scholastic, Inc., 1997.

Regan, Colm. *People of the Islands.* Austin, Tex.: Raintree Steck-Vaughn, 1998.

Waterlow, Julia. *Islands.* Austin, Tex.: Raintree Steck-Vaughn, 1995.

Index